Semiconductor Physics: A Formula Handbook

N.B. Singh

DEDICATION

To Nature,

I dedicate this book to you, the source of all life. You are my inspiration, my teacher, and my friend.

Thank you for teaching me about the beauty of the world around me. Thank you for showing me the power of the natural world. Thank you for giving me a sense of peace and tranquillity.

I promise to do my part to protect you and your many wonders. I will teach my children about the importance of conservation and sustainability. I will work to make the world a better place for all living things.

Thank you for everything, Nature.

With love,

N.B Singh

Contents

Preface

Welcome to "Semiconductor Physics: A Formula Handbook." This handbook is designed to serve as a comprehensive reference for professionals, researchers, and students delving into the intricate world of semiconductor physics.

Objective

The primary objective of this handbook is to provide a consolidated collection of essential formulas and concepts in semiconductor physics. From crystal structures to device modeling, each section is crafted to offer a quick and accessible reference for understanding and applying the fundamental principles.

Structure of the Handbook

The book is organized into several chapters, each focusing on a specific aspect of semiconductor physics. From introductory materials to advanced topics, readers will find a systematic arrangement of formulas, explanations, and practical examples.

Audience

This handbook is intended for a diverse audience, including engineers, physicists, and students engaged in semiconductor research, device design, or academic studies. The concise format aims to facilitate quick access to information, making it a valuable companion in research labs, classrooms, and professional settings.

Enjoy the Journey

As you navigate through the pages of "Semiconductor Physics: A Formula Handbook," we hope you find clarity in complex concepts and inspiration for your semiconductor endeavors.

Chapter 1

Introduction to

Semiconductor Physics

1.1 Basics of Semiconductor Materials

Semiconductor materials play a crucial role in electronic devices. Understanding their fundamental properties is essential for semiconductor physics. One of the key characteristics is the bandgap energy (E_g), which determines the material's conductivity.

The energy bandgap (E_g) is defined as the energy difference between the valence band and the conduction band. In a semiconductor, electrons in the valence band can move to the conduction band by absorbing energy equal to or greater than E_g. The formula for the energy bandgap is given by:

$$E_g = E_{\mathrm{CB}} - E_{\mathrm{VB}}$$

where E_{CB} is the energy of the conduction band and E_{VB} is the energy of the valence band.

Now, let's consider a numerical example. Suppose we have a semiconductor with a bandgap energy $E_g = 1.2\,\mathrm{eV}$. If an electron absorbs energy greater than

1

1.2 eV, it can move from the valence band to the conduction band, allowing the material to conduct electricity.

This fundamental property of semiconductors forms the basis for various electronic devices, such as diodes and transistors. As we delve deeper into semiconductor physics, we will explore how manipulating these materials leads to the development of advanced electronic components.

Semiconductor materials can be categorized into intrinsic and extrinsic types. Intrinsic semiconductors are pure materials without any added impurities, while extrinsic semiconductors contain impurities intentionally introduced during fabrication.

The carrier concentration in a semiconductor is a critical factor influencing its conductivity. In an intrinsic semiconductor at absolute zero temperature, the carrier concentration is primarily due to thermal excitation. As the temperature increases, more electron-hole pairs are generated, increasing the carrier concentration.

In contrast, extrinsic semiconductors can be either n-type or p-type based on the type of doping elements introduced. Doping with elements like phosphorus increases the electron concentration, resulting in an n-type semiconductor. On the other hand, doping with elements like boron increases the hole concentration, leading to a p-type semiconductor.

Understanding these basic concepts is foundational to semiconductor physics and device design. In the subsequent chapters, we will explore advanced topics, mathematical formulations, and practical applications of semiconductor physics.

1.2 Crystal Structure and Lattice Dynamics

The crystal structure of semiconductor materials significantly influences their electronic properties. Understanding crystal structures and lattice dynamics is crucial for comprehending semiconductor behavior. In this section, we'll explore the basics of crystal structures and their impact on semiconductor physics.

One fundamental concept is the Bravais lattice, which represents the repeated arrangement of atoms in a crystal. The lattice is characterized by three vectors (a_1, a_2, a_3) and three angles (α, β, γ). The lattice points form a periodic structure, and the crystal structure is determined by the arrangement of atoms within this lattice.

The crystal lattice's symmetry plays a vital role in the electronic band structure of semiconductors. For instance, in a simple cubic lattice, the Brillouin zone is a cube. In more complex structures, such as the face-centered cubic (FCC) or body-centered cubic (BCC) lattices, the Brillouin zone's shape varies.

The lattice dynamics of semiconductors involve the study of atomic vibrations within the crystal lattice. These vibrations, or phonons, affect thermal conductivity and electronic band structure. The Debye model provides an approximation for the phonon dispersion relation, connecting the phonon frequency (ω) to the wave vector (k).

$$\omega = v_s \cdot k$$

where v_s is the speed of sound in the material. This relation helps in understanding how thermal energy is transported through the crystal lattice.

Now, let's consider a numerical example. Suppose we have a semiconductor with a face-centered cubic lattice structure. The lattice constant a is $4\,\text{Å}$. The Brillouin zone's volume (V_{BZ}) for an FCC lattice is given by:

$$V_{\text{BZ}} = \left(\frac{\pi}{a}\right)^3$$

Substituting the given lattice constant:

$$V_{\text{BZ}} = \left(\frac{\pi}{4\,\text{Å}}\right)^3$$

Solving this, we find the Brillouin zone volume. Understanding such geometric and dynamic properties of crystal structures lays the foundation for more advanced semiconductor physics concepts.

Crystal defects, such as vacancies and interstitials, also impact semiconductor behavior. These defects can introduce energy levels within the bandgap, affecting carrier mobility and electronic properties.

1.3   Energy Bands in Solids

The concept of energy bands in solids is fundamental to understanding the electronic behavior of semiconductors. In this section, we will explore the formation of energy bands and their significance in the context of semiconductor physics.

In a solid, the arrangement of atoms leads to the creation of energy bands. The electronic band structure describes the distribution of energy levels for electrons in a material. The two primary types of energy bands are the valence band (VB) and the conduction band (CB). The energy gap between these bands is crucial and is known as the bandgap (E_g).

The energy bands are formed by the overlapping of atomic orbitals in the crystal lattice. In the valence band, electrons are tightly bound to atoms, while in the conduction band, electrons are free to move and contribute to electrical conductivity.

The energy dispersion relation in a crystal is given by the effective mass approximation:

$$E(k) = \frac{\hbar^2 k^2}{2m^*} + E_0$$

where $E(k)$ is the electron energy, $\hbar$ is the reduced Planck's constant, k is the wave vector, m^* is the effective mass, and E_0 is a constant term.

Now, let's consider a numerical example. Suppose we have a semiconductor with an effective mass $m^* = 0.1m_0$, where m_0 is the free electron mass. The energy dispersion relation for this semiconductor is given by:

$$E(k) = \frac{\hbar^2 k^2}{2 \times 0.1m_0} + E_0$$

This relation helps us understand how the energy of electrons changes with their wave vector in the crystal lattice.

The bandgap (E_g) is a critical parameter in semiconductor devices. For example, a material with a larger bandgap is an insulator, while a smaller bandgap material is a semiconductor. The energy bands and bandgap play a crucial role in determining the electrical conductivity and optical properties of semiconductors.

The movement of electrons between the valence and conduction bands is central to the operation of semiconductor devices. In the absence of external factors, electrons in the valence band remain in a low-energy state, and the material acts as an insulator. However, when energy is supplied, electrons can move to the conduction band, allowing for electrical conduction.

The energy-momentum relationship also impacts carrier mobility, which is essential for designing high-performance semiconductor devices. The effective mass of carriers influences their response to an applied electric field, affecting how quickly they can move through the material.

1.4   Intrinsic and Extrinsic Semiconductors

Semiconductors can be broadly categorized into intrinsic and extrinsic types based on their purity and conductivity characteristics. In this section, we will delve into the distinctions between these semiconductor types and their impact on electronic behavior.

An intrinsic semiconductor is a pure semiconductor without intentional doping. Silicon (Si) and germanium (Ge) are examples of intrinsic semiconductors. In these materials, electrons in the valence band can be excited to the conduction band due to thermal energy, creating electron-hole pairs. The carrier concentration in intrinsic semiconductors is solely dependent on temperature.

The carrier concentration (n_i) in an intrinsic semiconductor is given by the intrinsic carrier concentration formula:

$$n_i = \sqrt{N_c \cdot N_v} \cdot e^{-\frac{E_g}{2kT}}$$

where N_c and N_v are the effective densities of states in the conduction and valence bands, respectively, E_g is the energy bandgap, k is the Boltzmann constant, and T is the absolute temperature.

Now, let's consider a numerical example. For silicon with an energy bandgap $E_g = 1.1\,\mathrm{eV}$ at room temperature ($T = 300\,\mathrm{K}$), the intrinsic carrier concentration is calculated as:

$$n_i = \sqrt{N_c \cdot N_v} \cdot e^{-\frac{1.1}{2 \times 8.617 \times 10^{-5} \times 300}}$$

This example illustrates how intrinsic carrier concentration varies with temperature and bandgap.

Extrinsic semiconductors are intentionally doped to modify their electrical properties. Doping introduces impurity atoms into the crystal lattice, creating additional energy levels within the bandgap. There are two primary types of extrinsic semiconductors: n-type and p-type.

In n-type semiconductors, such as silicon doped with phosphorus, the additional electrons from the dopant contribute to the conduction band. The electron concentration (n) in n-type semiconductors is given by:

$$n = N_d$$

where N_d is the donor concentration.

On the other hand, in p-type semiconductors, such as silicon doped with boron, the dopant creates holes in the valence band. The hole concentration (p) in p-type semiconductors is given by:

$$p = N_a$$

where N_a is the acceptor concentration.

The total charge neutrality condition in a semiconductor is given by:

$$n \cdot p = n_i^2$$

This equation ensures that the product of electron and hole concentrations equals the square of the intrinsic carrier concentration.

In practical applications, controlling the type and concentration of dopants allows engineers to tailor semiconductor properties for specific device requirements. For instance, the controlled doping of silicon forms the basis for the development of diodes, transistors, and other electronic components.

Chapter 2

Semiconductor Devices and Technology

2.1 Diodes and Rectifiers

Diodes are semiconductor devices crucial for rectification and signal processing in electronic circuits. In this section, we will explore the physics of diodes, their characteristics, and their applications in rectifiers.

A diode is a two-terminal semiconductor device with a p-n junction. The current-voltage relationship in a diode is described by the Shockley diode equation:

$$I = I_s \left(e^{\frac{V}{nV_T}} - 1 \right)$$

where I is the diode current, I_s is the reverse saturation current, V is the voltage across the diode, n is the ideality factor, and V_T is the thermal voltage given by $V_T = \frac{kT}{q}$ (where k is the Boltzmann constant, T is the temperature, and q is the elementary charge).

Now, let's consider a numerical example. Suppose we have a silicon diode with a reverse saturation current $I_s = 1\,\text{nA}$, an ideality factor $n = 1.5$, and a temperature of 300 K. If the diode is forward-biased with a voltage of 0.7 V, we can calculate the diode current using the Shockley diode equation.

$$I = 1\,\text{nA}\left(e^{\frac{0.7}{1.5\times0.026}} - 1\right)$$

This example demonstrates how to use the Shockley diode equation to determine the current through a diode under specific biasing conditions.

Rectifiers are circuits that convert alternating current (AC) to direct current (DC) by utilizing diodes. The most common rectifier is the bridge rectifier, consisting of four diodes arranged in a bridge configuration. The output voltage (V_{out}) of a bridge rectifier can be approximated by:

$$V_{\text{out}} = V_{\text{peak}} - 2V_{\text{diode}}$$

where V_{peak} is the peak voltage of the AC input and V_{diode} is the voltage drop across one diode during conduction.

Consider a bridge rectifier with a peak AC voltage of 10 V. If each diode has a voltage drop of 0.7 V during conduction, the DC output voltage is given by:

$$V_{\text{out}} = 10 - 2 \times 0.7$$

This formula illustrates the relationship between the AC input and DC output in a bridge rectifier.

The efficiency (η) of a rectifier is defined as the ratio of DC power output to AC power input. For a bridge rectifier, the efficiency is given by:

$$\eta = \frac{V_{\text{out,DC}}^2}{2V_{\text{peak,AC}}^2}$$

This formula quantifies the effectiveness of the rectification process.

Diodes and rectifiers play a crucial role in power supply circuits and electronic devices. Understanding their characteristics and behavior is essential for designing efficient and reliable circuits.

2.2   Bipolar Junction Transistors (BJTs)

Bipolar Junction Transistors (BJTs) are essential semiconductor devices with applications ranging from amplifiers to switches. In this section, we will delve into the physics of BJTs, exploring their operating principles and key characteristics.

A BJT comprises three layers: emitter (E), base (B), and collector (C). The current-voltage relationship in a BJT is described by the Shockley diode equation for each junction:

$$I_{\text{BE}} = I_{\text{S}} \left(e^{\frac{V_{\text{BE}}}{nV_T}} - 1 \right)$$

$$I_{\text{BC}} = I_{\text{S}} \left(e^{\frac{V_{\text{BC}}}{nV_T}} - 1 \right)$$

where I_{BE} and I_{BC} are the base-emitter and base-collector currents, V_{BE} and V_{BC} are the base-emitter and base-collector voltages, I_{S} is the reverse saturation current, n is the ideality factor, and V_T is the thermal voltage.

Consider a numerical example with an NPN transistor having $I_{\text{S}} = 10\,\text{nA}$, $n = 1.2$, and $V_T = 26\,\text{mV}$ at room temperature. If the base-emitter voltage is $V_{\text{BE}} = 0.7\,\text{V}$, we can calculate the base-emitter current I_{BE}:

$$I_{\text{BE}} = 10\,\text{nA} \left(e^{\frac{0.7}{1.2 \times 0.026}} - 1 \right)$$

Understanding the Shockley equation is crucial for determining currents in a BJT.

BJTs operate in three regions: cutoff, active, and saturation. The transistor amplification factor (β) is the ratio of collector current (I_C) to base current (I_B) in the active region:

$$\beta = \frac{I_C}{I_B}$$

The collector current in the active region is related to the base-emitter voltage (V_{BE}) by the Ebers-Moll model:

$$I_C = I_S \left(e^{\frac{V_{BE}}{nV_T}} - 1 \right)$$

Understanding these relationships is essential for designing and analyzing transistor circuits.

In BJT amplifier circuits, the small-signal common-emitter voltage gain (A_v) is given by:

$$A_v = -\beta \cdot R_L$$

where R_L is the load resistance.

Consider a common-emitter amplifier with a transistor having $\beta = 100$ and $R_L = 5\,\text{k}\omega$. The voltage gain is:

$$A_v = -100 \times 5\,\text{k}\omega$$

This example illustrates how to calculate the voltage gain in a common-emitter amplifier.

Bipolar Junction Transistors are fundamental to modern electronics, enabling a wide range of circuit functionalities. The Shockley diode equation and Ebers-Moll model provide a foundation for understanding BJT behavior, and their application in amplifier circuits is crucial for signal processing in electronic devices.

2.3   Metal-Oxide-Semiconductor Field-Effect Transistors (MOSFETs)

Metal-Oxide-Semiconductor Field-Effect Transistors (MOSFETs) are fundamental semiconductor devices extensively used in integrated circuits for amplification and digital switching. In this section, we will explore the physics of MOSFETs, their characteristics, and operational principles.

MOSFETs consist of a metal gate separated from the semiconductor by a thin insulating layer, typically silicon dioxide. The device's operation relies on the control of the channel between the source (S) and drain (D) terminals by the voltage applied to the gate (G) terminal.

The MOSFET threshold voltage (V_{th}) is a critical parameter determining the onset of channel formation. The threshold voltage is influenced by factors such as the work function of the metal gate (Φ_{ms}), semiconductor electron affinity (χ), and oxide semiconductor interface charge (Q_{it}). The threshold voltage is given by:

$$V_{th} = \Phi_{ms} + \chi + \frac{Q_{it}}{C_{ox}}$$

where C_{ox} is the capacitance per unit area of the oxide.

Now, let's consider a numerical example. For a MOSFET with $\Phi_{ms} = 4.1\,\text{eV}$, $\chi = 4.2\,\text{eV}$, and $Q_{it} = 1 \times 10^{11}\,\text{cm}^{-2}$, with $C_{ox} = 3.9 \times 10^{-8}\,\text{F/cm}^2$, the threshold voltage is calculated as:

$$V_{th} = 4.1 + 4.2 + \frac{1 \times 10^{11}}{3.9 \times 10^{-8}}$$

This example illustrates the determination of the MOSFET threshold voltage based on relevant parameters.

The MOSFET drain current (I_D) in the saturation region is given by the quadratic equation:

$$I_D = \frac{1}{2} \mu_n C_{ox} \frac{W}{L} (V_{GS} - V_{th})^2$$

where μ_n is the electron mobility, W is the channel width, and L is the channel length.

Consider a MOSFET with $\mu_n = 600\,\text{cm}^2/\text{V-s}$, $W = 10\,\mu\text{m}$, $L = 1\,\mu\text{m}$, $V_{GS} - V_{th} = 0.2\,\text{V}$. The drain current is:

$$I_D = \frac{1}{2} \times 600 \times 10^{-8} \times (0.2)^2$$

This formula demonstrates the calculation of drain current in the saturation region.

MOSFETs are characterized by their transconductance (g_m), which relates small changes in the gate-source voltage (V_{GS}) to changes in the drain current. The transconductance is given by:

$$g_m = \mu_n C_{ox} \frac{W}{L} (V_{GS} - V_{th})$$

Understanding transconductance is crucial for designing MOSFET-based amplifiers.

In MOSFET amplifier circuits, the small-signal voltage gain (A_v) is given by:

$$A_v = -g_m \times R_D$$

where R_D is the drain resistor.

Consider a common-source amplifier with a MOSFET having $g_m = 1.2\,\text{mS}$ and $R_D = 5\,\text{k}\omega$. The voltage gain is:

$$A_v = -1.2 \times 5$$

This formula illustrates the calculation of voltage gain in a common-source amplifier.

2.4 Junction Field-Effect Transistors (JFETs)

Junction Field-Effect Transistors (JFETs) are semiconductor devices that rely on the control of current flow between the source (S) and drain (D) terminals by a voltage applied to the gate (G) terminal. In this section, we will explore the physics of JFETs, their characteristics, and operational principles.

JFETs come in two main types: N-channel (with an N-type channel) and P-channel (with a P-type channel). The drain current (I_D) in a JFET is governed by the square law equation:

$$I_D = I_{DSS} \left(1 - \frac{V_{GS}}{V_P}\right)^2$$

where I_{DSS} is the saturation drain current, V_{GS} is the gate-source voltage, and V_{P} is the pinch-off voltage.

Now, let's consider a numerical example. For an N-channel JFET with $I_{\text{DSS}} = 5\,\text{mA}$ and $V_{\text{P}} = -4\,\text{V}$, if the gate-source voltage is $V_{\text{GS}} = -1\,\text{V}$, we can calculate the drain current using the square law equation:

$$I_{\text{D}} = 5 \times 10^{-3} \left(1 - \frac{-1}{-4}\right)^2$$

This example demonstrates the use of the square law equation to determine the drain current in an N-channel JFET.

JFETs are voltage-controlled devices, and their transconductance (g_m) is a crucial parameter. The transconductance is given by:

$$g_m = \frac{2I_{\text{DSS}}}{|V_{\text{P}}|} \sqrt{1 - \frac{V_{\text{GS}}}{V_{\text{P}}}}$$

Understanding transconductance is essential for designing JFET-based amplifiers.

Consider a common-source amplifier with an N-channel JFET having $I_{\text{DSS}} = 2\,\text{mA}$ and $V_{\text{P}} = -6\,\text{V}$. If the gate-source voltage is $V_{\text{GS}} = -2\,\text{V}$, we can calculate the transconductance using the formula:

$$g_m = \frac{2 \times 2 \times 10^{-3}}{|-6|} \sqrt{1 - \frac{-2}{-6}}$$

This formula illustrates the determination of transconductance in a common-source amplifier.

In JFET circuits, the small-signal voltage gain (A_{v}) is given by:

$$A_{\text{v}} = -g_m \times R_{\text{D}}$$

where R_{D} is the drain resistor.

Consider a JFET amplifier with $g_m = 2\,\text{mS}$ and $R_{\text{D}} = 4\,\text{k}\omega$. The voltage gain is:

$$A_{\text{v}} = -2 \times 4$$

This formula illustrates the calculation of voltage gain in a JFET amplifier.

2.5   Semiconductor Fabrication Technology

Semiconductor fabrication technology plays a pivotal role in the manufacturing of electronic devices. This section delves into the key processes and considerations involved in the fabrication of semiconductor devices.

One of the fundamental processes in semiconductor fabrication is photolithography. The minimum feature size (L) that can be achieved in photolithography is given by the Rayleigh criterion:

$$L = \frac{k \cdot \lambda}{\text{NA}}$$

where k is a constant, λ is the wavelength of the light source, and NA is the numerical aperture. This formula provides insights into the resolution limits of the photolithography process.

Consider a photolithography system with a wavelength $\lambda = 193\,\text{nm}$ and a numerical aperture NA of 0.75. Using the Rayleigh criterion, we can calculate the minimum feature size L:

$$L = \frac{k \cdot 193}{0.75}$$

This example illustrates how the Rayleigh criterion is applied in determining the minimum feature size.

Another critical aspect of semiconductor fabrication is the doping of semiconductor materials. The surface concentration (N_s) of dopants in a semiconductor material is related to the dopant density (N_d) and the depletion region width (W) by the equation:

$$N_s = \frac{N_d}{2} \sqrt{\frac{q \cdot \varepsilon \cdot \text{NA}}{2 \cdot \phi}}$$

where q is the elementary charge, ε is the permittivity of the material, and ϕ is the surface potential.

For example, consider a silicon substrate with a dopant density $N_d = 1 \times 10^{18}$ cm^{-3}, a depletion region width $W = 0.1\,\mu$m, and assuming silicon has a permittivity $\varepsilon = 11.7$. Using the formula, we can determine the surface concentration N_s:

$$N_s = \frac{1 \times 10^{18}}{2} \sqrt{\frac{1.6 \times 10^{-19} \cdot 11.7 \cdot 0.1 \times 10^{-4}}{2 \times \phi}}$$

This example showcases how the surface concentration is calculated based on doping and depletion region parameters.

In the process of ion implantation, the dopant concentration profile can be described by the Pearson IV distribution:

$$C(x) = C_0 \exp\left(-\frac{2x^2}{w^2}\right)$$

where $C(x)$ is the dopant concentration at a depth x, C_0 is the peak concentration, and w is the standard deviation. This distribution characterizes the spatial distribution of dopants after ion implantation.

Consider an ion implantation process with a peak concentration $C_0 = 1 \times 10^{16}$ cm^{-3} and a standard deviation $w = 50$ nm. Using the Pearson IV distribution, we can determine the dopant concentration at various depths x.

The deposition of thin films is a crucial step in semiconductor fabrication. The thickness (d) of a deposited film can be calculated using the deposition rate (R_{dep}) and the deposition time (t):

$$d = R_{\text{dep}} \cdot t$$

For instance, if the deposition rate is $R_{\text{dep}} = 5$ nm/min and the deposition time is $t = 30$ min, the film thickness is:

$$d = 5 \cdot 30$$

This example demonstrates how the film thickness is determined based on deposition rate and time.

In semiconductor fabrication, thermal processes are employed to anneal and activate dopants. The diffusion length (L_d) of dopants in a semiconductor material during thermal annealing is given by:

$$L_d = \sqrt{\frac{D \cdot t}{\pi}}$$

where D is the diffusion coefficient and t is the annealing time.

Consider an annealing process with a diffusion coefficient $D = 1 \times 10^{-12} \, \text{cm}^2/\text{s}$ and an annealing time $t = 10 \, \text{min}$. Using the formula, we can calculate the diffusion length L_d:

$$L_d = \sqrt{\frac{1 \times 10^{-12} \cdot 600}{\pi}}$$

This example illustrates how the diffusion length is determined during thermal annealing.

Chapter 3

Carrier Transport Phenomena

3.1 Charge Carriers in Semiconductors

Understanding the behavior of charge carriers in semiconductors is fundamental to grasping the principles of semiconductor physics. This section explores the characteristics and transport phenomena associated with charge carriers in semiconductor materials.

The carrier concentration (n) in an intrinsic semiconductor can be determined using the intrinsic carrier concentration (n_i), which is a function of temperature (T):

$$n = n_i^2 \exp\left(\frac{E_g}{2kT}\right)$$

where E_g is the energy bandgap, k is Boltzmann's constant, and T is the absolute temperature.

Consider a silicon crystal with an energy bandgap $E_g = 1.12\,\text{eV}$ at room temperature $(T = 300\,\text{K})$. The intrinsic carrier concentration n_i is given by:

$$n_i = \sqrt{N_c \cdot N_v} \exp\left(-\frac{E_g}{2kT}\right)$$

where N_c and N_v are the effective densities of states in the conduction and valence bands. Using the provided values, we can calculate the intrinsic carrier concentration n_i and subsequently, the carrier concentration n in the intrinsic silicon crystal.

The mobility of charge carriers in a semiconductor (μ) is a crucial parameter affecting the rate of carrier transport. The mobility is related to the carrier relaxation time (τ) and the charge (q) by:

$$\mu = \frac{q \cdot \tau}{m^*}$$

where m^* is the effective mass of the charge carrier.

Consider an N-type silicon semiconductor with an electron mobility $\mu_n = 1400\,\mathrm{cm^2/V\text{-}s}$ and a carrier concentration $n = 1 \times 10^{16}\,\mathrm{cm^{-3}}$. Using the formula, we can determine the electron relaxation time τ_n:

$$\tau_n = \frac{\mu_n \cdot m_n^*}{q}$$

This example illustrates how to calculate the electron relaxation time in an N-type semiconductor.

The drift velocity (v_d) of charge carriers in a semiconductor subjected to an electric field (E) is given by:

$$v_d = \mu \cdot E$$

where E is the electric field.

Consider a P-type silicon semiconductor with a hole mobility $\mu_p = 500\,\mathrm{cm^2/V\text{-}s}$ and an electric field $E = 100\,\mathrm{V/cm}$. Using the formula, we can calculate the drift velocity v_{d_p} of holes:

$$v_{d_p} = \mu_p \cdot E$$

This example demonstrates the determination of drift velocity in a P-type semiconductor under an applied electric field.

In semiconductor devices, the current (I) flowing through a device is related to the charge carrier concentration and drift velocity by the formula:

$$I = q \cdot A \cdot n \cdot v_d$$

where A is the cross-sectional area of the device.

Consider a silicon diode with a cross-sectional area $A = 1 \times 10^{-4} \, \text{cm}^2$, a carrier concentration $n = 1 \times 10^{17} \, \text{cm}^{-3}$, and a drift velocity $v_d = 0.1 \, \text{cm/s}$. Using the formula, we can calculate the current flowing through the diode:

$$I = q \cdot A \cdot n \cdot v_d$$

This example showcases the calculation of current in a semiconductor device.

3.2 Carrier Drift and Mobility

The movement of charge carriers in a semiconductor under the influence of an electric field is a crucial aspect of semiconductor physics. This section explores the concepts of carrier drift and mobility, providing insights into their behavior.

The drift velocity (v_d) of charge carriers in a semiconductor subjected to an electric field (E) is given by the formula:

$$v_d = \mu \cdot E$$

where E is the electric field and μ is the mobility of the charge carriers.

Consider an N-type silicon semiconductor with an electron mobility $\mu_n = 1400 \, \text{cm}^2/\text{V-s}$ and an electric field $E = 100 \, \text{V/cm}$. Using the formula, we can calculate the drift velocity v_{d_n} of electrons:

$$v_{d_n} = \mu_n \cdot E$$

This example illustrates the determination of drift velocity in an N-type semiconductor under an applied electric field.

Similarly, for a P-type semiconductor with hole mobility μ_p, the drift velocity v_{d_p} of holes is given by:

$$v_{d_p} = \mu_p \cdot E$$

Understanding drift velocity is crucial for analyzing the motion of charge carriers in semiconductor devices.

The mobility (μ) of charge carriers in a semiconductor is influenced by factors such as temperature (T) and scattering mechanisms. For electrons in an N-type semiconductor, the mobility is given by the relation:

$$\mu_n = \frac{e\tau_n}{m_n^*}$$

where e is the elementary charge, τ_n is the electron relaxation time, and m_n^* is the effective mass of electrons.

Consider an N-type silicon semiconductor with an electron relaxation time $\tau_n = 1 \times 10^{-13}$ s and an effective mass $m_n^* = 0.26\, m_0$ (m_0 is the free electron mass). Using the formula, we can determine the electron mobility μ_n:

$$\mu_n = \frac{e \cdot 1 \times 10^{-13}}{0.26 \cdot m_0}$$

This example showcases the calculation of electron mobility in an N-type semiconductor.

Similarly, for holes in a P-type semiconductor, the hole mobility (μ_p) is given by:

$$\mu_p = \frac{e\tau_p}{m_p^*}$$

where τ_p is the hole relaxation time and m_p^* is the effective mass of holes.

In semiconductor devices, the current (I) flowing through a device can be expressed in terms of the charge carrier mobility, carrier concentration, and applied electric field:

$$I = q \cdot A \cdot n \cdot \mu \cdot E$$

where q is the elementary charge, A is the cross-sectional area of the device, n is the carrier concentration, μ is the mobility, and E is the electric field.

Consider a silicon diode with a cross-sectional area $A = 1 \times 10^{-4}\,\mathrm{cm}^2$, a carrier concentration $n = 1 \times 10^{16}\,\mathrm{cm}^{-3}$, a mobility $\mu = 1400\,\mathrm{cm}^2/\mathrm{V\text{-}s}$, and an electric field $E = 10\,\mathrm{V/cm}$. Using the formula, we can calculate the current flowing through the diode:

$$I = q \cdot A \cdot n \cdot \mu \cdot E$$

This example illustrates the practical application of the current formula in semiconductor devices.

3.3 Carrier Diffusion

Carrier diffusion is a significant process influencing the movement of charge carriers in semiconductors. This section explores the fundamental principles of carrier diffusion and provides insights into its behavior.

The diffusion current (J_{diff}) in a semiconductor is described by Fick's first law of diffusion:

$$J_{\mathrm{diff}} = -q \cdot D \cdot \frac{dn}{dx}$$

where q is the elementary charge, D is the diffusion coefficient, dn is the change in carrier concentration, and dx is the change in position.

Consider an N-type semiconductor with a carrier concentration gradient $\frac{dn}{dx} = 1 \times 10^{14}\,\mathrm{cm}^{-4}$ and a diffusion coefficient $D_n = 25\,\mathrm{cm}^2/\mathrm{s}$. Using Fick's law, we can calculate the diffusion current:

$$J_{\mathrm{diff}} = -q \cdot 25 \cdot 1 \times 10^{14}$$

This example illustrates the application of Fick's law to determine the diffusion current in an N-type semiconductor.

Similarly, for P-type semiconductors, the diffusion current is given by:

$$J_{\text{diff}} = q \cdot D \cdot \frac{dp}{dx}$$

where dp is the change in hole concentration.

In a semiconductor with both electron and hole diffusion, the total diffusion current is the sum of the electron and hole diffusion currents:

$$J_{\text{total}} = J_{\text{diff, n}} + J_{\text{diff, p}}$$

Understanding carrier diffusion is crucial for analyzing the behavior of charge carriers in different semiconductor regions.

The diffusion length (L_{diff}) characterizes the spatial extent over which carriers diffuse before recombining. For a semiconductor with a constant carrier concentration and a constant diffusion coefficient, the diffusion length is given by:

$$L_{\text{diff}} = \sqrt{D \cdot \tau}$$

where D is the diffusion coefficient and τ is the carrier lifetime.

Consider a silicon semiconductor with a diffusion coefficient $D = 20\,\text{cm}^2/\text{s}$ and a carrier lifetime $\tau = 1 \times 10^{-6}\,\text{s}$. Using the formula, we can calculate the diffusion length:

$$L_{\text{diff}} = \sqrt{20 \cdot 1 \times 10^{-6}}$$

This example demonstrates how to determine the diffusion length in a semiconductor.

In semiconductor devices, the diffusion current can contribute to the total current alongside the drift current. The total current (I_{total}) is the sum of the diffusion current and drift current:

$$I_{\text{total}} = q \cdot \left(D \cdot \frac{dn}{dx} \cdot E + \mu \cdot n \cdot E\right)$$

where E is the electric field.

Consider a silicon diode with an electric field $E = 500\,\text{V/cm}$, a carrier concentration gradient $\frac{dn}{dx} = 5 \times 10^{15}\,\text{cm}^{-4}$, a diffusion coefficient $D = 15\,\text{cm}^2/\text{s}$, and a mobility $\mu = 1200\,\text{cm}^2/\text{V-s}$. Using the formula, we can calculate the total current:

$$I_{\text{total}} = q \cdot (15 \cdot 5 \times 10^{15} + 1200 \cdot 5 \times 10^{15})$$

This example illustrates the calculation of the total current considering both diffusion and drift.

3.4 Generation and Recombination of Carriers

The generation and recombination of carriers in semiconductors play a pivotal role in determining the overall behavior of semiconductor devices. This section explores the mechanisms behind carrier generation and recombination and provides insights into their impact on semiconductor physics.

The rate of carrier generation (G) due to thermal processes in a semiconductor material can be described by the equation:

$$G = B(T) \cdot e^{-\frac{E_g}{kT}}$$

where $B(T)$ is the temperature-dependent generation constant, E_g is the energy bandgap, k is Boltzmann's constant, and T is the absolute temperature.

Consider a silicon material with an energy bandgap $E_g = 1.12\,\text{eV}$ and a temperature $T = 300\,\text{K}$. The generation constant $B(T)$ can be expressed as:

$$B(T) = A \cdot T^n$$

where A and n are constants. For this example, let's assume $A = 1 \times 10^{16}\,\mathrm{cm}^{-3} \cdot \mathrm{s}^{-1}$ and $n = 2$. Using the given values, we can calculate the rate of carrier generation G:

$$G = 1 \times 10^{16} \cdot 300^2 \cdot e^{-\frac{1.12}{8.617 \times 10^{-5} \cdot 300}}$$

This example illustrates the calculation of the rate of carrier generation in silicon at room temperature.

Carrier recombination (R) in semiconductors can occur through various mechanisms. For direct band-to-band recombination, the recombination rate is given by:

$$R = C \cdot n \cdot p$$

where C is the recombination constant, n is the electron concentration, and p is the hole concentration.

Consider a silicon semiconductor with an electron concentration $n = 1 \times 10^{16}\,\mathrm{cm}^{-3}$ and a hole concentration $p = 1 \times 10^{15}\,\mathrm{cm}^{-3}$. Assuming $C = 1 \times 10^{-11}\,\mathrm{cm}^3/\mathrm{s}$, we can calculate the recombination rate R:

$$R = 1 \times 10^{-11} \cdot 1 \times 10^{16} \cdot 1 \times 10^{15}$$

This example demonstrates the determination of the recombination rate through direct band-to-band recombination.

Auger recombination is another mechanism where one electron and one hole recombine, resulting in the generation of a new electron-hole pair. The rate of Auger recombination (R_{Auger}) is given by:

$$R_{\mathrm{Auger}} = C_{\mathrm{Auger}} \cdot n \cdot p^2$$

where C_{Auger} is the Auger recombination constant.

For example, if $C_{\mathrm{Auger}} = 1 \times 10^{-31}\,\mathrm{cm}^6/\mathrm{s}$, $n = 5 \times 10^{15}\,\mathrm{cm}^{-3}$, and $p = 1 \times 10^{15}\,\mathrm{cm}^{-3}$, we can calculate the Auger recombination rate R_{Auger}:

$$R_{\text{Auger}} = 1 \times 10^{-31} \cdot 5 \times 10^{15} \cdot (1 \times 10^{15})^2$$

This example showcases the calculation of the Auger recombination rate in a semiconductor.

In semiconductor devices, the net rate of carrier generation or recombination influences the carrier concentration and device performance. The continuity equation, representing the rate of change of carrier concentration, is given by:

$$\frac{dn}{dt} = G - R$$

This equation describes how carriers are generated and recombine in the semiconductor material.

Chapter 4

Semiconductor

Optoelectronics

4.1   Optical Properties of Semiconductors

The optical properties of semiconductors play a crucial role in the field of optoelectronics, influencing the interaction of semiconductors with light. This section explores the key optical properties and their significance in semiconductor
physics.

The absorption coefficient (α) characterizes the rate at which a semiconductor absorbs light and is related to the intensity of incident light (I) and
transmitted light (I_t) by Beer-Lambert's law:

$$I_t = I \cdot e^{-\alpha x}$$

where x is the distance traveled through the semiconductor.

Consider a semiconductor with an absorption coefficient $\alpha = 1\,\mathrm{cm}^{-1}$ and a
thickness $x = 0.1\,\mathrm{cm}$. Using Beer-Lambert's law, we can calculate the transmitted light intensity I_t:

$$I_t = I \cdot e^{-1 \cdot 0.1}$$

This example illustrates the application of Beer-Lambert's law to determine the transmitted light intensity through a semiconductor.

The absorption coefficient can also be expressed in terms of the energy of the incident photons ($h\nu$), the speed of light (c), and the semiconductor's energy bandgap (E_g):

$$\alpha = \frac{A}{\sqrt{E_g - h\nu}}$$

where A is a material-dependent constant.

For example, for a semiconductor with an energy bandgap $E_g = 1.5\,\text{eV}$ and incident photons of energy $h\nu = 2\,\text{eV}$, we can calculate the absorption coefficient:

$$\alpha = \frac{A}{\sqrt{1.5 - 2}}$$

This formula demonstrates the relationship between the absorption coefficient and the energy of incident photons.

The refractive index (n) of a semiconductor influences the speed of light within the material and is related to the speed of light in a vacuum (c) by:

$$n = \frac{c}{v}$$

where v is the phase velocity of light in the semiconductor.

Consider a semiconductor with a refractive index $n = 3$. Using the formula, we can calculate the phase velocity v of light in the semiconductor:

$$v = \frac{c}{n}$$

This example illustrates the determination of the phase velocity based on the refractive index.

The reflectance (R) of a semiconductor surface is related to the refractive index by the Fresnel equations. For light incident at an angle (θ), the reflectance is given by:

$$R = \left(\frac{n_1 \cos(\theta) - n_2 \sqrt{1 - \left(\frac{n_1}{n_2} \sin(\theta) \right)^2}}{n_1 \cos(\theta) + n_2 \sqrt{1 - \left(\frac{n_1}{n_2} \sin(\theta) \right)^2}} \right)^2$$

where n_1 and n_2 are the refractive indices of the two media.

For example, if light is incident from air $(n_1 = 1)$ onto a semiconductor surface with $n_2 = 3$ at an angle $\theta = 30°$, we can calculate the reflectance:

$$R = \left(\frac{1 \cos(30°) - 3 \sqrt{1 - \left(\frac{1}{3} \sin(30°) \right)^2}}{1 \cos(30°) + 3 \sqrt{1 - \left(\frac{1}{3} \sin(30°) \right)^2}} \right)^2$$

This example demonstrates the calculation of reflectance using the Fresnel equations.

4.2 Light Emitting Diodes (LEDs)

Light Emitting Diodes (LEDs) are semiconductor devices that emit light when current flows through them. This section explores the principles behind LED operation, the factors influencing their performance, and relevant formulas.

The basic working principle of an LED involves the recombination of charge carriers in the semiconductor material, leading to the emission of photons. The energy of the emitted photons is related to the energy bandgap (E_g) of the semiconductor material by Planck's equation:

$$E = h\nu = E_g$$

where E is the energy of the photon, h is Planck's constant, and ν is the frequency of the emitted light.

Consider an LED with a semiconductor material characterized by an energy bandgap $E_g = 2\,\text{eV}$. Using Planck's equation, we can calculate the energy and frequency of the emitted photons:

$$E = h\nu = 2\,\text{eV}$$

This example illustrates the determination of the energy and frequency of emitted light from an LED.

The wavelength (λ) of the emitted light is related to the frequency by the speed of light (c):

$$\lambda = \frac{c}{\nu}$$

Consider an LED emitting light with a frequency $\nu = 5 \times 10^{14}\,\text{Hz}$. Using the formula, we can calculate the wavelength of the emitted light:

$$\lambda = \frac{3 \times 10^8}{5 \times 10^{14}}$$

This example demonstrates the calculation of the wavelength of light emitted by an LED.

The radiative recombination rate (R_{rad}) in an LED, representing the rate of emission of photons, is given by:

$$R_{\text{rad}} = B \cdot n \cdot p$$

where B is the radiative recombination constant, n is the electron concentration, and p is the hole concentration.

Consider an LED with an electron concentration $n = 1 \times 10^{17}\,\text{cm}^{-3}$ and a hole concentration $p = 1 \times 10^{16}\,\text{cm}^{-3}$. Assuming $B = 1 \times 10^{-10}\,\text{cm}^3/\text{s}$, we can calculate the radiative recombination rate:

$$R_{\text{rad}} = 1 \times 10^{-10} \cdot 1 \times 10^{17} \cdot 1 \times 10^{16}$$

This example showcases the determination of the radiative recombination rate in an LED.

The external quantum efficiency (η_{ext}) of an LED, representing the efficiency of converting injected charge carriers into emitted photons, is given by:

$$\eta_{\text{ext}} = \frac{R_{\text{rad}}}{e \cdot I}$$

where e is the elementary charge and I is the current flowing through the LED.

Consider an LED with a current $I = 20\,\text{mA}$. Using the formula, we can calculate the external quantum efficiency:

$$\eta_{\text{ext}} = \frac{1 \times 10^{-10} \cdot 1 \times 10^{17} \cdot 1 \times 10^{16}}{1.6 \times 10^{-19} \cdot 0.02}$$

This example illustrates the calculation of the external quantum efficiency of an LED.

4.3 Semiconductor Lasers

Semiconductor lasers, commonly known as laser diodes, are essential components in optoelectronics, enabling the generation of coherent and monochromatic light. This section explores the principles governing the operation of semiconductor lasers and relevant formulas.

The threshold current (I_{th}) is a crucial parameter for semiconductor lasers and represents the minimum current required for laser action. It can be related to the modal gain (G) and the cavity loss (α) by:

$$I_{\text{th}} = \frac{\alpha}{G}$$

Consider a semiconductor laser with a cavity loss $\alpha = 10\,\text{cm}^{-1}$ and a modal gain $G = 2 \times 10^{1}5\,\text{cm}^{-1}$. Using the formula, we can calculate the threshold current:

$$I_{th} = \frac{10}{2 \times 10^{15}}$$

This example illustrates the determination of the threshold current for a semiconductor laser.

The laser output power (P_{out}) of a semiconductor laser is related to the injected current (I), slope efficiency (η), and threshold current by:

$$P_{out} = \eta \cdot (I - I_{th})$$

Consider a semiconductor laser with a threshold current $I_{th} = 20\,\text{mA}$, an injected current $I = 40\,\text{mA}$, and a slope efficiency $\eta = 0.3$. Using the formula, we can calculate the laser output power:

$$P_{out} = 0.3 \cdot (40 - 20)$$

This example showcases the calculation of the laser output power for a given semiconductor laser.

The external quantum efficiency (η_{ext}) of a semiconductor laser, representing the efficiency of converting injected electrical power into laser power, is given by:

$$\eta_{ext} = \frac{P_{out}}{e \cdot I}$$

where e is the elementary charge.

Consider a semiconductor laser with a laser output power $P_{out} = 2\,\text{mW}$ and an injected current $I = 50\,\text{mA}$. Using the formula, we can calculate the external quantum efficiency:

$$\eta_{ext} = \frac{2 \times 10^{-3}}{1.6 \times 10^{-19} \cdot 0.05}$$

This example demonstrates the calculation of the external quantum efficiency for a semiconductor laser.

The linewidth (Δv) of a semiconductor laser is related to the spontaneous emission factor (α_s), speed of light (c), and the effective refractive index (n_{eff}) by:

$$\Delta v = \frac{\alpha_s \cdot c}{2 \cdot \alpha \cdot n_{eff}}$$

Consider a semiconductor laser with a spontaneous emission factor $\alpha_s = 2 \times 10^{12} \, \text{Hz}$, an effective refractive index $n_{eff} = 3$. Using the formula, we can calculate the linewidth:

$$\Delta v = \frac{2 \times 10^{12} \cdot 3 \times 10^8}{2 \cdot \alpha \cdot 3}$$

This example illustrates the calculation of the linewidth for a semiconductor laser.

4.4 Photodetectors and Solar Cells

Photodetectors and solar cells are semiconductor devices that play a crucial role in converting light energy into electrical signals or electrical power. This section explores the principles behind photodetectors and solar cells, providing relevant formulas and examples.

The responsivity (R) of a photodetector, representing its ability to convert incident optical power into electrical current, is given by:

$$R = \frac{I_{ph}}{P_{in}}$$

where I_{ph} is the photocurrent and P_{in} is the incident optical power.

Consider a photodetector with a photocurrent $I_{ph} = 5 \, \mu\text{A}$ and an incident optical power $P_{in} = 1 \, \text{mW}$. Using the formula, we can calculate the responsivity:

$$R = \frac{5 \times 10^{-6}}{1 \times 10^{-3}}$$

This example illustrates the determination of the responsivity for a photodetector.

The quantum efficiency (η_{QE}) of a photodetector, representing the efficiency of converting incident photons into photocurrent, is given by:

$$\eta_{QE} = \frac{e \cdot I_{ph}}{h \cdot v}$$

where e is the elementary charge, I_{ph} is the photocurrent, h is Planck's constant, and v is the optical frequency.

Consider a photodetector with a photocurrent $I_{ph} = 10\,\mu A$ and an optical frequency $v = 5 \times 10^{14}\,\text{Hz}$. Using the formula, we can calculate the quantum efficiency:

$$\eta_{QE} = \frac{1.6 \times 10^{-19} \cdot 10 \times 10^{-6}}{6.63 \times 10^{-34} \cdot 5 \times 10^{14}}$$

This example demonstrates the calculation of the quantum efficiency for a photodetector.

The open-circuit voltage (V_{oc}) of a solar cell, representing the maximum voltage across its terminals in the absence of an external load, is related to the thermal voltage (V_T) and the number of electron-hole pairs generated (G) by:

$$V_{oc} = G \cdot V_T$$

Consider a solar cell with $G = 2 \times 10^{16}\,\text{cm}^{-3}$ and $V_T = 26\,\text{mV}$ at a certain temperature. Using the formula, we can calculate the open-circuit voltage:

$$V_{oc} = 2 \times 10^{16} \cdot 26 \times 10^{-3}$$

This example illustrates the determination of the open-circuit voltage for a solar cell.

The short-circuit current (I_{sc}) of a solar cell, representing the maximum current when the terminals are short-circuited, is related to the charge of the electron (e), the number of electron-hole pairs generated (G), and the absorption coefficient (α) by:

$$I_{sc} = e \cdot G \cdot \frac{d}{\alpha}$$

where d is the thickness of the solar cell.

Consider a solar cell with $G = 1 \times 10^{17}\,\text{cm}^{-3}$, $d = 100\,\mu\text{m}$, and $\alpha = 5 \times 10^3\,\text{cm}^{-1}$. Using the formula, we can calculate the short-circuit current:

$$I_{\text{sc}} = 1.6 \times 10^{-19} \cdot 1 \times 10^{17} \cdot \frac{100 \times 10^{-4}}{5 \times 10^3}$$

This example demonstrates the calculation of the short-circuit current for a solar cell.

Chapter 5

Advanced Semiconductor Topics

5.1 Heterojunctions and Quantum Wells

Heterojunctions and quantum wells are advanced semiconductor structures that play a pivotal role in modern electronic and optoelectronic devices. This section delves into the principles governing heterojunctions and quantum wells, providing relevant formulas and examples.

The band lineup at a heterojunction, where two different semiconductor materials meet, is crucial for understanding charge carrier behavior. The energy band offset (ΔE_c for conduction band and ΔE_v for valence band) between the two materials is a key parameter. The electron affinity (χ) and the difference in electron affinities ($\Delta\chi$) contribute to the band offset:

$$\Delta E_c = \chi_{\text{material 2}} - \chi_{\text{material 1}} + \Delta\chi$$

Consider a heterojunction between GaAs ($\chi = 4.07\,\text{eV}$) and AlAs ($\chi = 4.23\,\text{eV}$) with $\Delta\chi = 0.1\,\text{eV}$. Using the formula, we can calculate the conduction band offset:

$$\Delta E_c = 4.23 - 4.07 + 0.1$$

This example illustrates the calculation of the conduction band offset for a heterojunction.

The quantum well confinement energy (E_{conf}) in a quantum well structure is related to the effective mass (m^*) and the width of the well (a):

$$E_{conf} = \frac{\hbar^2 \pi^2}{2m^* a^2}$$

Consider a quantum well with an effective mass $m^* = 0.07 m_0$ (where m_0 is the free electron mass) and a well width $a = 10\,\text{nm}$. Using the formula, we can calculate the quantum well confinement energy:

$$E_{conf} = \frac{(6.626 \times 10^{-34}/(2\pi))^2}{2 \times 0.07 \times 9.1 \times 10^{-31} \times (10 \times 10^{-9})^2}$$

This example showcases the calculation of the confinement energy for a quantum well.

The density of states ($D(E)$) in a quantum well is related to the effective mass and the well width. For a two-dimensional electron gas, $D(E)$ is given by:

$$D(E) = \frac{1}{\pi \hbar^2} \sqrt{\frac{2m^*}{\hbar^2}(E - E_{conf})}$$

Consider a quantum well with an effective mass $m^* = 0.1 m_0$, a well width $a = 15\,\text{nm}$, and a confinement energy $E_{conf} = 50\,\text{meV}$. Using the formula, we can calculate the density of states:

$$D(E) = \frac{1}{\pi (6.626 \times 10^{-34})^2} \sqrt{\frac{2 \times 0.1 \times 9.1 \times 10^{-31}}{(E - 50 \times 10^{-3})}}$$

This example demonstrates the calculation of the density of states for a quantum well.

Heterojunctions and quantum wells find applications in devices such as lasers, transistors, and photodetectors. The precise control of electronic states and energy levels in these structures enables enhanced device performance. Formulas and examples provided in this section offer insights into the fundamental aspects of heterojunctions and quantum wells in advanced semiconductor topics.

5.2 Semiconductor Nanostructures

Semiconductor nanostructures, including quantum dots and nanowires, exhibit unique electronic and optical properties due to their reduced dimensions. This section explores the principles governing semiconductor nanostructures, providing relevant formulas and examples.

The energy levels of a quantum dot are quantized due to confinement, and the energy separation between levels (E_{spacing}) is related to the effective mass (m^*) and the size of the quantum dot (R):

$$E_{\text{spacing}} = \frac{\pi^2 \hbar^2}{2m^* R^2}$$

Consider a quantum dot with an effective mass $m^* = 0.08 m_0$ (where m_0 is the free electron mass) and a radius $R = 5\,\text{nm}$. Using the formula, we can calculate the energy spacing:

$$E_{\text{spacing}} = \frac{\pi^2 (6.626 \times 10^{-34}/(2\pi))^2}{2 \times 0.08 \times 9.1 \times 10^{-31} \times (5 \times 10^{-9})^2}$$

This example illustrates the calculation of the energy spacing for a quantum dot.

The density of states ($D(E)$) in a nanowire is related to the effective mass and the cross-sectional area. For a one-dimensional electron gas, $D(E)$ is given by:

$$D(E) = \frac{1}{\pi \hbar^2} \sqrt{\frac{2m^*}{\hbar^2}} (E - E_{\text{conf}})$$

Consider a nanowire with an effective mass $m^* = 0.1 m_0$ and a cross-sectional area $A = 10\,\text{nm}^2$. Using the formula, we can calculate the density of states:

$$D(E) = \frac{1}{\pi(6.626 \times 10^{-34})^2} \sqrt{\frac{2 \times 0.1 \times 9.1 \times 10^{-31}}{(E - E_{\text{conf}})}}$$

This example demonstrates the calculation of the density of states for a nanowire.

The tunneling probability (T) through a potential barrier in a nanoscale device is given by the formula:

$$T = e^{-2\kappa d}$$

where κ is the decay constant and d is the thickness of the barrier. The decay constant is related to the energy of the particle (E), the barrier height (V_0), and the effective mass (m^*):

$$\kappa = \frac{\sqrt{2m^*(V_0 - E)}}{\hbar}$$

Consider a nanoscale device with a potential barrier height $V_0 = 0.2\,\text{eV}$, an energy of the particle $E = 0.1\,\text{eV}$, an effective mass $m^* = 0.15m_0$, and a barrier thickness $d = 2\,\text{nm}$. Using the formulas, we can calculate the decay constant and the tunneling probability:

$$\kappa = \frac{\sqrt{2 \times 0.15 \times 9.1 \times 10^{-31}(0.2 - 0.1)}}{6.626 \times 10^{-34}}$$

$$T = e^{-2 \times \frac{\sqrt{2 \times 0.15 \times 9.1 \times 10^{-31}(0.2-0.1)} \times 2 \times 10^{-9}}{6.626 \times 10^{-34}}}$$

This example showcases the calculation of the tunneling probability through a nanoscale potential barrier.

Semiconductor nanostructures play a crucial role in the development of next-generation electronic and optoelectronic devices. Understanding their unique properties and behaviors is essential for designing and optimizing nanoscale devices. Formulas and examples provided in this section offer insights into the fundamental aspects of semiconductor nanostructures in advanced semiconductor topics.

5.3   Semiconductor Device Modeling

Semiconductor device modeling involves the mathematical representation of device behavior for analysis and simulation. This section explores the key aspects of semiconductor device modeling, providing relevant formulas and examples.

The current-voltage $(I - V)$ characteristics of a semiconductor device, such as a diode, can be described using the Shockley diode equation:

$$I = I_{\text{sat}} \left(e^{\frac{V}{nV_{\text{th}}}} - 1 \right)$$

where I_{sat} is the saturation current, V is the voltage across the diode, n is the ideality factor, and V_{th} is the thermal voltage.

Consider a diode with $I_{\text{sat}} = 1\,\mu\text{A}$, $n = 1.5$, and $V_{\text{th}} = 26\,\text{mV}$. Using the Shockley diode equation, we can analyze the $I - V$ characteristics for different values of V:

$$I(V) = 1 \times 10^{-6} \left(e^{\frac{V}{1.5 \times 26 \times 10^{-3}}} - 1 \right)$$

This example illustrates the application of the Shockley diode equation for modeling diode behavior.

The small-signal parameters of a transistor, such as the transconductance (g_m) and output conductance (g_{d}), can be expressed using the hybrid-π transistor model:

$$i_{\text{out}} = g_m v_{\text{in}} + g_{\text{d}} v_{\text{out}}$$

where i_{out} is the output current, v_{in} is the input voltage, and v_{out} is the output voltage.

Consider a transistor with $g_m = 50\,\text{mS}$ and $g_{\text{d}} = 5\,\text{mS}$. Using the hybrid-$\pi$ model, we can analyze the small-signal behavior of the transistor:

$$i_{\text{out}} = 50 \times 10^{-3} v_{\text{in}} + 5 \times 10^{-3} v_{\text{out}}$$

This example demonstrates the use of the hybrid-π model for small-signal analysis in transistor modeling.

The SPICE (Simulation Program with Integrated Circuit Emphasis) simulation tool is widely used for semiconductor device modeling. SPICE models incorporate various parameters and equations to simulate the behavior of electronic circuits accurately.

Consider a simple circuit with a resistor, capacitor, and voltage source in SPICE. The SPICE netlist for this circuit may look like:

R1 1 2 1k

C1 2 0 1uF

V1 1 0 DC 5V

This SPICE netlist models a circuit with a 1k-ohm resistor, a 1uF capacitor, and a 5V DC voltage source.

Semiconductor device modeling is essential for the design and optimization of electronic circuits. Various mathematical models and simulation tools, such as SPICE, enable engineers to predict and analyze the behavior of semiconductor devices accurately.

5.4 Power Semiconductor Devices

Power semiconductor devices are essential components in electronic systems for controlling and managing electrical power. This section explores the principles and characteristics of power semiconductor devices, providing relevant formulas and examples.

The on-state voltage drop (V_{on}) across a power diode can be described by the diode forward voltage drop formula:

$$V_{on} = V_{th} \ln \left(1 + \frac{I_{on}}{I_{sat}} \right)$$

where V_{th} is the thermal voltage, I_{on} is the on-state current, and I_{sat} is the saturation current.

Consider a power diode with $I_{on} = 50\,\text{A}$, $I_{sat} = 5\,\text{mA}$, and $V_{th} = 26\,\text{mV}$. Using the diode forward voltage drop formula, we can calculate V_{on}:

$$V_{on} = 26 \times 10^{-3} \ln \left(1 + \frac{50}{5 \times 10^{-3}} \right)$$

This example illustrates the calculation of the on-state voltage drop for a power diode.

The switching loss (P_{switch}) in a power semiconductor device, such as a metal-oxide-semiconductor field-effect transistor (MOSFET), can be determined by:

$$P_{switch} = \frac{1}{2} C_{oss} V_{in}^2 f_{sw}$$

where C_{oss} is the output capacitance, V_{in} is the input voltage, and f_{sw} is the switching frequency.

Consider a MOSFET with $C_{oss} = 500\,\text{pF}$, $V_{in} = 100\,\text{V}$, and $f_{sw} = 100\,\text{kHz}$. Using the switching loss formula, we can calculate P_{switch}:

$$P_{switch} = \frac{1}{2} \times 500 \times 10^{-12} \times (100)^2 \times 100 \times 10^3$$

This example demonstrates the calculation of switching loss in a power MOSFET.

The thermal resistance (R_{th}) of a power semiconductor device, such as a insulated-gate bipolar transistor (IGBT), is given by:

$$R_{th} = \frac{T_{junction} - T_{ambient}}{P_{diss}}$$

where $T_{junction}$ is the junction temperature, $T_{ambient}$ is the ambient temperature, and P_{diss} is the power dissipation.

Consider an IGBT with $T_{\text{junction}} = 125\,°C$, $T_{\text{ambient}} = 25\,°C$, and $P_{\text{diss}} = 10\,W$. Using the thermal resistance formula, we can calculate R_{th}:

$$R_{\text{th}} = \frac{125 - 25}{10}$$

This example showcases the calculation of thermal resistance in a power IGBT.

Power semiconductor devices are crucial in various applications, including power electronics and electric power systems. Understanding their characteristics and behavior is essential for designing efficient and reliable power electronic systems. Formulas and examples provided in this section offer insights into the fundamental aspects of power semiconductor devices in advanced semiconductor topics.

Chapter 6

Semiconductor Physics Formulas and Equations

6.1 Fundamental Semiconductor Equations

Semiconductor physics relies on fundamental equations that describe the behavior of carriers and materials. This section explores key equations governing semiconductor physics, providing relevant formulas and examples.

The continuity equation expresses the conservation of charge carriers in a semiconductor:

$$\frac{\partial n}{\partial t} + \nabla \cdot \mathbf{J}_n = G_n - R_n$$

where n is the electron concentration, $\mathbf{J}_n$ is the electron current density, G_n is the generation rate, and R_n is the recombination rate.

Consider a semiconductor with a generation rate $G_n = 1 \times 10^{15}\,\mathrm{cm^{-3}s^{-1}}$ and a recombination rate $R_n = 5 \times 10^{14}\,\mathrm{cm^{-3}s^{-1}}$. Using the continuity equation, we can analyze the change in electron concentration over time.

$$\frac{\partial n}{\partial t} + \nabla \cdot \mathbf{J}_n = 1 \times 10^{15} - 5 \times 10^{14}$$

This example illustrates the application of the continuity equation in semiconductor physics.

The drift-diffusion equation combines the effects of carrier drift and diffusion:

$$\frac{\partial n}{\partial t} + \nabla \cdot \mathbf{J}_n = \mu_n \mathbf{E} \cdot n + D_n \nabla^2 n - R_n$$

where μ_n is the electron mobility, $\mathbf{E}$ is the electric field, and D_n is the electron diffusion coefficient.

Consider a semiconductor with an electron mobility $\mu_n = 1500 \, \text{cm}^2/\text{Vs}$ and a diffusion coefficient $D_n = 30 \, \text{cm}^2/\text{s}$. Using the drift-diffusion equation, we can analyze the combined effects of drift, diffusion, and recombination.

$$\frac{\partial n}{\partial t} + \nabla \cdot \mathbf{J}_n = 1500 \times \mathbf{E} \cdot n + 30 \times \nabla^2 n - R_n$$

This example demonstrates the application of the drift-diffusion equation in semiconductor physics.

The Shockley equation describes the current through a p-n junction diode:

$$I = I_{\text{sat}} \left(e^{\frac{V}{nV_{\text{th}}}} - 1 \right)$$

where I_{sat} is the saturation current, V is the voltage across the diode, n is the ideality factor, and V_{th} is the thermal voltage.

Consider a p-n junction diode with $I_{\text{sat}} = 1 \, \mu \, \text{A}$, $n = 1.2$, and $V_{\text{th}} = 26 \, \text{mV}$. Using the Shockley equation, we can analyze the diode current for different values of V.

$$I(V) = 1 \times 10^{-6} \left(e^{\frac{V}{1.2 \times 26 \times 10^{-3}}} - 1 \right)$$

This example illustrates the application of the Shockley equation in semiconductor diode analysis.

Understanding these fundamental semiconductor equations is crucial for predicting and analyzing the behavior of semiconductor devices. The examples provided showcase their practical applications in semiconductor physics.

6.2 Device Performance Equations

Device performance in semiconductor physics is often characterized by specific equations that quantify key parameters. This section explores equations related to device performance, providing relevant formulas and examples.

The carrier lifetime (τ) in a semiconductor is a critical parameter affecting device performance and is related to the recombination rate (R):

$$\tau = \frac{1}{R}$$

Consider a semiconductor material with a recombination rate $R = 1 \times 10^{14}\,\mathrm{cm}^{-3}\mathrm{s}^{-1}$. Using the formula, we can calculate the carrier lifetime:

$$\tau = \frac{1}{1 \times 10^{14}}$$

This example illustrates the determination of carrier lifetime based on recombination rate.

The external quantum efficiency (EQE) of a photodetector measures the efficiency of converting incident photons into photocurrent and is given by:

$$EQE = \frac{\text{Number of Photons Converted to Photocurrent}}{\text{Number of Incident Photons}}$$

Consider a photodetector that converts 80 out of 100 incident photons into photocurrent. Using the formula, we can calculate the external quantum efficiency:

$$EQE = \frac{80}{100} = 0.8$$

This example demonstrates the calculation of external quantum efficiency for a photodetector.

The power conversion efficiency (η) of a solar cell is a crucial performance metric and is determined by the ratio of the maximum power point ($P_{\max}$) to the incident solar power (P_{sun}):

$$\eta = \frac{P_{\text{max}}}{P_{\text{sun}}}$$

Consider a solar cell with $P_{\text{max}} = 4\,\text{W}$ and $P_{\text{sun}} = 10\,\text{W}$. Using the power conversion efficiency formula, we can calculate η:

$$\eta = \frac{4}{10} = 0.4$$

This example showcases the calculation of power conversion efficiency for a solar cell.

The charge carrier mobility (μ) in a semiconductor, a key parameter for transistor performance, is given by the relationship between carrier drift velocity (v_d) and electric field (E):

$$\mu = \frac{v_d}{E}$$

Consider a semiconductor with a carrier drift velocity $v_d = 0.1\,\text{cm/s}$ under an electric field $E = 1\,\text{V/cm}$. Using the mobility formula, we can calculate μ:

$$\mu = \frac{0.1}{1} = 0.1\,\text{cm}^2/\text{Vs}$$

This example demonstrates the determination of charge carrier mobility based on drift velocity and electric field.

Understanding these device performance equations is crucial for evaluating and optimizing the efficiency of semiconductor devices. The examples provided showcase their practical applications in assessing the performance of various semiconductor devices.

6.3 Numerical Simulation Techniques

Numerical simulation techniques play a crucial role in understanding and predicting the behavior of semiconductor devices. This section explores the fundamental equations and methods used in numerical simulations, providing relevant formulas and examples.

The drift-diffusion equations, which describe the transport of carriers in semiconductors, can be solved numerically to simulate device behavior. The equations for electron transport (n) are given by:

$$\frac{\partial n}{\partial t} + \nabla \cdot \mathbf{J}_n = \mu_n \mathbf{E} \cdot n + D_n \nabla^2 n - R_n$$

where $\mathbf{J}_n$ is the electron current density, μ_n is the electron mobility, $\mathbf{E}$ is the electric field, D_n is the electron diffusion coefficient, and R_n is the recombination rate.

Consider a numerical simulation of carrier transport in a semiconductor device. By discretizing the spatial and temporal domains, numerical methods like finite difference or finite element can be employed to solve these equations and predict the electron concentration distribution over time and space.

The Poisson equation, which describes the electrostatics in semiconductors, is another essential equation in numerical simulations. For a one-dimensional case, the Poisson equation is given by:

$$\frac{d}{dx}\left(\varepsilon(x)\frac{d\phi}{dx}\right) = -q(p - n + N_D - N_A)$$

where $\varepsilon(x)$ is the permittivity, ϕ is the electrostatic potential, q is the elementary charge, p and n are the hole and electron concentrations, and N_D and N_A are the donor and acceptor concentrations.

In a numerical simulation of a semiconductor device, solving the Poisson equation helps obtain the spatial distribution of the electrostatic potential. This is crucial for understanding the electric field distribution within the device.

Numerical techniques such as finite element methods or boundary element methods are commonly employed to solve these partial differential equations in semiconductor device simulations. These methods discretize the device structure and solve the equations iteratively to capture the complex interactions among carriers, electric fields, and material properties.

Consider a simulation of a p-n junction device using finite element methods. The simulation can provide insights into the electric field distribution, charge carrier concentrations, and current-voltage characteristics of the device under different operating conditions.

In summary, numerical simulation techniques are indispensable for gaining a deeper understanding of semiconductor device behavior. They allow researchers and engineers to predict and optimize device performance by solving the complex set of equations governing carrier transport, electrostatics, and other physical phenomena.

Chapter 7

Appendix: Mathematical Background

7.1 Vector and Matrix Algebra

Vector and matrix algebra form the foundation of mathematical tools used in semiconductor physics. This section provides an overview of essential concepts, formulas, and examples.

7.1.1 Vector Algebra

In semiconductor physics, vectors are often used to represent physical quantities like electric field, current density, and carrier velocity. A vector $\mathbf{A}$ in three-dimensional space is typically denoted as:

$$\mathbf{A} = A_x \mathbf{i} + A_y \mathbf{j} + A_z \mathbf{k}$$

where A_x, A_y, and A_z are the vector components along the x, y, and z axes, respectively.

Scalar multiplication of a vector $\mathbf{A}$ by a scalar c is given by:

$$c\mathbf{A} = cA_x\mathbf{i} + cA_y\mathbf{j} + cA_z\mathbf{k}$$

Vector addition involves adding corresponding components:

$$\mathbf{A} + \mathbf{B} = (A_x + B_x)\mathbf{i} + (A_y + B_y)\mathbf{j} + (A_z + B_z)\mathbf{k}$$

Example 1: Consider two vectors $\mathbf{A} = 2\mathbf{i} + 3\mathbf{j} - \mathbf{k}$ and $\mathbf{B} = -\mathbf{i} + 4\mathbf{j} + 2\mathbf{k}$. The sum $\mathbf{A} + \mathbf{B}$ is calculated as:

$$\mathbf{A} + \mathbf{B} = (2 - 1)\mathbf{i} + (3 + 4)\mathbf{j} + (-1 + 2)\mathbf{k} = \mathbf{i} + 7\mathbf{j} + \mathbf{k}$$

7.1.2 Matrix Algebra

Matrices are used to represent linear transformations, such as those encountered in semiconductor device simulations. A matrix $\mathbf{M}$ with dimensions $m \times n$ has entries M_{ij}, where i represents the row index and j represents the column index.

Matrix addition involves adding corresponding entries:

$$\mathbf{C} = \mathbf{A} + \mathbf{B} \implies C_{ij} = A_{ij} + B_{ij}$$

Matrix multiplication is defined as:

$$\mathbf{C} = \mathbf{A} \cdot \mathbf{B} \implies C_{ij} = \sum_{k=1}^{n} A_{ik} \cdot B_{kj}$$

Example 2: Consider matrices

$$\mathbf{A} = \begin{bmatrix} 1 & 2 \\ 3 & 4 \end{bmatrix}, \quad \mathbf{B} = \begin{bmatrix} 5 & 6 \\ 7 & 8 \end{bmatrix}$$

The product $\mathbf{C} = \mathbf{A} \cdot \mathbf{B}$ is calculated as:

$$\mathbf{C} = \begin{bmatrix} 1 \cdot 5 + 2 \cdot 7 & 1 \cdot 6 + 2 \cdot 8 \\ 3 \cdot 5 + 4 \cdot 7 & 3 \cdot 6 + 4 \cdot 8 \end{bmatrix} = \begin{bmatrix} 19 & 22 \\ 43 & 50 \end{bmatrix}$$

7.1.3 Determinant and Inverse

The determinant of a square matrix $\mathbf{A}$ is denoted as $\det(\mathbf{A})$ and is calculated using a recursive formula. The inverse of a square matrix $\mathbf{A}$, denoted as $\mathbf{A}^{-1}$, satisfies $\mathbf{A} \cdot \mathbf{A}^{-1} = \mathbf{A}^{-1} \cdot \mathbf{A} = \mathbf{I}$, where $\mathbf{I}$ is the identity matrix.

Example 3: For a 2x2 matrix $\mathbf{A} = \begin{bmatrix} a & b \\ c & d \end{bmatrix}$, the determinant is $\det(\mathbf{A}) = ad - bc$ and the inverse is given by

$$\mathbf{A}^{-1} = \frac{1}{ad - bc} \begin{bmatrix} d & -b \\ -c & a \end{bmatrix}$$

These fundamental concepts of vector and matrix algebra are crucial for understanding mathematical formulations in semiconductor physics and device simulations.

7.2 Differential Equations in Semiconductor Physics

Differential equations play a fundamental role in describing the behavior of carriers, electric fields, and other physical quantities in semiconductor physics. This section provides an overview of differential equations commonly encountered in semiconductor physics, along with relevant formulas and examples.

7.2.1 Carrier Transport Equations

The transport of carriers in semiconductors is often described by drift-diffusion equations. For electrons (n) and holes (p), the continuity equation is given by:

$$\frac{\partial n}{\partial t} + \nabla \cdot \mathbf{J}_n = G_n - R_n$$

$$\frac{\partial p}{\partial t} - \nabla \cdot \mathbf{J}_p = G_p - R_p$$

where $\mathbf{J}_n$ and $\mathbf{J}_p$ are the electron and hole current densities, G_n and G_p are the generation rates, and R_n and R_p are the recombination rates.

Example 1: Consider a one-dimensional semiconductor with an initial electron concentration $n_0(x) = 1 \times 10^{16} \, \text{cm}^{-3}$ and a constant generation rate $G_n = 1 \times 10^{14} \, \text{cm}^{-3}\text{s}^{-1}$. The solution for the electron concentration as a function of time and position can be obtained by solving the drift-diffusion equation numerically.

7.2.2   Poisson's Equation

Poisson's equation describes the electrostatics in semiconductors and is given by:

$$\nabla \cdot (\varepsilon(x)\nabla\phi) = -q(p - n + N_D - N_A)$$

where $\varepsilon(x)$ is the permittivity, ϕ is the electrostatic potential, q is the elementary charge, p and n are the hole and electron concentrations, and N_D and N_A are the donor and acceptor concentrations.

Example 2: Consider a semiconductor with a uniform acceptor concentration $N_A = 1 \times 10^{17} \, \text{cm}^{-3}$ and a uniform donor concentration $N_D = 1 \times 10^{16} \, \text{cm}^{-3}$. The solution for the electrostatic potential $\phi(x)$ can be obtained by solving Poisson's equation for the given boundary conditions.

7.2.3   Diffusion Equation

The diffusion equation describes the spatial variation of carrier concentrations due to diffusion:

$$\frac{\partial n}{\partial t} = D_n \nabla^2 n + G_n - R_n$$

$$\frac{\partial p}{\partial t} = D_p \nabla^2 p + G_p - R_p$$

where D_n and D_p are the electron and hole diffusion coefficients.

Example 3: Consider a semiconductor initially in thermal equilibrium with a step change in the concentration of minority carriers. The solution for the spatial and temporal evolution of carrier concentrations can be obtained by solving the diffusion equation.

In semiconductor device simulations, these differential equations are typically solved using numerical techniques such as finite difference methods or finite element methods. The solutions provide insights into the dynamic behavior of carriers and electric fields within semiconductor devices.

7.3 Statistical Mechanics Concepts

Statistical mechanics provides the theoretical framework to understand the behavior of a large number of particles in a system. In semiconductor physics, statistical mechanics is crucial for describing the statistical distribution of electrons and holes. This section introduces key concepts and formulas related to statistical mechanics in the context of semiconductors.

7.3.1 Fermi-Dirac Distribution Function

The distribution of electrons in a semiconductor is described by the Fermi-Dirac distribution function, denoted as $f(E)$:

$$f(E) = \frac{1}{\exp\left(\frac{E-E_F}{kT}\right) + 1}$$

where: - E is the energy of a state, - E_F is the Fermi energy, - k is the Boltzmann constant, - T is the temperature.

The Fermi-Dirac distribution function indicates the probability of finding an electron at a given energy level.

Example 1: Consider a semiconductor at room temperature ($T = 300\,\mathrm{K}$) with a Fermi energy E_F of $0.3\,\mathrm{eV}$. Calculate the probability of finding an electron at an energy level $E = 0.1\,\mathrm{eV}$.

$$f(0.1\,\mathrm{eV}) = \frac{1}{\exp\left(\frac{0.1-0.3}{0.0259}\right) + 1}$$

7.3.2 Density of States

The density of states $(D(E))$ describes the number of energy states per unit energy interval at a given energy level E:

$$D(E) = \frac{dN}{dE}$$

In a semiconductor, the density of states depends on the effective mass (m^*) of charge carriers and is often expressed as:

$$D(E) = \frac{1}{2\pi^2} \left(\frac{2m^*}{\hbar^2} \right)^{3/2} \sqrt{E}$$

where: - m^* is the effective mass of charge carriers, - $\hbar$ is the reduced Planck constant.

Example 2: For an electron in silicon ($m^* \approx 0.26m_e$, where m_e is the electron rest mass), calculate the density of states at an energy level $E = 0.5\,\text{eV}$.

$$D(0.5\,\text{eV}) = \frac{1}{2\pi^2} \left(\frac{2 \times 0.26m_e}{\hbar^2} \right)^{3/2} \sqrt{0.5}$$

7.3.3 Maxwell-Boltzmann Distribution Function

The Maxwell-Boltzmann distribution function describes the statistical distribution of particles in a gas at equilibrium:

$$f(v) = 4\pi \left(\frac{m}{2\pi kT} \right)^{3/2} v^2 \exp\left(-\frac{mv^2}{2kT} \right)$$

In semiconductor physics, it is often used to describe the distribution of carrier velocities.

Example 3: For electrons in a semiconductor at $T = 300\,\text{K}$, calculate the probability density of finding an electron with a velocity of $v = 10^5\,\text{m/s}$.

$$f(10^5\,\text{m/s}) = 4\pi \left(\frac{m}{2\pi kT} \right)^{3/2} (10^5)^2 \exp\left(-\frac{m(10^5)^2}{2kT} \right)$$

In summary, statistical mechanics concepts such as the Fermi-Dirac distribution, density of states, and Maxwell-Boltzmann distribution are essential for understanding the statistical behavior of charge carriers in semiconductors.